my itty-bitty bio

Jonas Salk

Cherry Lake Press

Published in the United States of America by Cherry Lake Publishing
Ann Arbor, Michigan
www.cherrylakepublishing.com

Reading Adviser: Beth Walker Gambro, MS, Ed., Reading Consultant, Yorkville, IL
Illustrator: Leo Trinidad

Photo Credits: FDR Presidential Library & Museum via Flickr, 5; © Dziurek / Shutterstock, 7; © ASSOCIATED PRESS, 9, 15; © ZUMA Press, Inc. / Alamy Stock Photo, 11; © RBM Vintage Images / Alamy Stock Photo, 13, 17, 22; © Photo 12 / Alamy Stock Photo, 19, 23; © Yousuf Karsh / Yousuf Karsh Estate, 21

Cherry Lake Press is an imprint of Cherry Lake Publishing Group

Library of Congress Cataloging-in-Publication Data has been filed and is available at catalog.loc.gov.

Printed in the United States of America

table of contents

About the author: Meeg Pincus is the award-winning author of 30 nonfiction picture books. She has worked for newspapers, book publishers, universities—and, years ago, at Disneyland. Meeg lives with her family in California, where she loves to write, read, make art, and sing.

About the illustrator: Leo Trinidad is a NY Times bestselling comic book artist, illustrator, and animator from Costa Rica. For more than 12 years he's been creating content for children's books and TV shows. Leo created the first animated series ever produced in Central America, and founded Rocket Cartoons, one of the most successful animation studios in Latin America. He is also the 2018 winner of the Central American Graphic Novel contest.

my story

I was born in New York City. It was 1914. A **virus** called polio hit when I was 2. Many people got sick and died.

My family did not get it.

Our **Jewish faith** says to "heal the world." I wanted to do that.

I worked hard in school.

I started college at age 15.
Then I went to **medical school**.

I chose to study viruses.

What do you like to study?

Polio came back every summer.
Young children got it most.
Pools and playgrounds closed.

I made a **vow** to stop it!

Scientists worked slowly on a **vaccine** to stop polio.

I wanted to make one quickly, my own way.

I made a polio vaccine in my **lab**. I believed it would work.

I tested it on myself and my family.

How does it feel to
believe in yourself?

My vaccine seemed to work. We needed to test it on more people.

More than 1 million children tested it.

It worked! The world cheered. Soon, we stopped polio in most countries.

I received many honors.

POLIO
VACCINE
WYETH
RUSH!
POLIO
VACCINE
WYETH
RUSH!
POLIO
VACCINE
WYETH
RUSH!
POLIO
VACCINE
POLIO
VACCINE
WYETH
AA

I gave away my polio vaccine for free. Later, I built a big lab to study viruses. Scientists still work there today.

They try to heal the world.

What would you like to ask me?

timeline

1955

1900

Born
1914

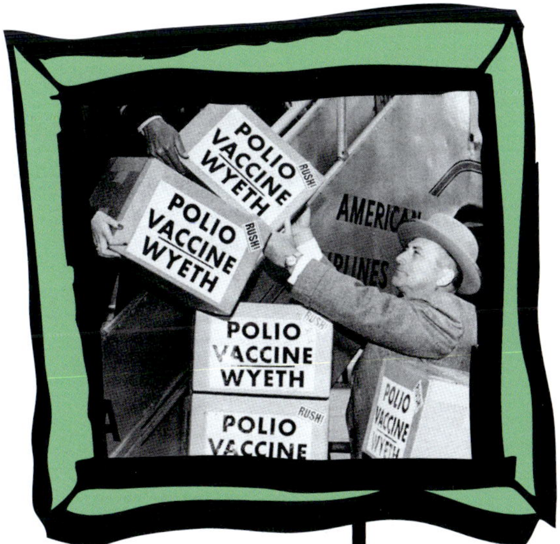

1961

2000

Died
1995

glossary & index

glossary

Jewish faith (JOO-ish FAYTH) the religion of Judaism, from the ancient Hebrews

lab (LAB) short for laboratory; a room or building where scientific experiments are done

medical school (MEH-dih-kuhl SKOOL) a college for people who want to be doctors

vaccine (VAK-zeen) a substance created to protect the body from a specific disease

virus (VIE-ruhs) a tiny, unseen agent that causes disease

vow (VOW) a serious promise

index